FACT FINDERS
KNOW THE DIFFERENCE!

FISH OR MARINE MAMMAL?

EZRA E. KNOPP

I0807209

PowerKiDS press.

Published in 2026 by The Rosen Publishing Group, Inc.
2544 Clinton Street, Buffalo, NY 14224

Copyright © 2026 by The Rosen Publishing Group, Inc.

All rights reserved. No part of this book may be reproduced in any form without permission in writing from the publisher, except by a reviewer.

First Edition

Editor: Greg Roza
Book Design: Michael Flynn

Photo Credits: Cover (fish) George Trumpeter/Shutterstock.com; cover (dolphin) Muhammadphotoes/Shutterstock.com; p. 5 (whale shark) mc_pongsatorn/Shutterstock.com; p. 5 (humpback whale) Wirestock Creators/Shutterstock.com; p. 6 Javidestock/Shutterstock.com; p. 7 Utku Cirak Photos/Shutterstock.com; p. 8 The Law of Adventures/Shutterstock.com; p. 9 ART_JA/Shutterstock.com; p. 10 Wirestock Creators/Shutterstock.com; p. 11 William2023/Shutterstock.com; p. 12 Risto Raunio/Shutterstock.com; p. 13 Ivan Derkach/Shutterstock.com; p. 14 slowmotiongli/Shutterstock.com; p. 15 UW.ART/Shutterstock.com; p. 16 Christina Dutkowski/Shutterstock.com; p. 17 outdoorsman/Shutterstock.com; p. 18 Caleb Foster/Shutterstock.com; p. 19 (sailfish) A Cotton Photo/Shutterstock.com; p. 19 (pufferfish) Miroslav Halama/Shutterstock.com; p. 21 (manatees) Natalie11345/Shutterstock.com; p. 21 (piranha) Tatiana Belova/Shutterstock.com.

Library of Congress Cataloging-in-Publication Data

Names: Knopp, Ezra E. author
Title: Fish or marine mammal? / Ezra E. Knopp.
Description: Buffalo, NY : PowerKids Press, [2026] | Series: Fact finders : know the difference! | Includes bibliographical references and index. | Audience term: juvenile
Identifiers: LCCN 2024061399 (print) | LCCN 2024061400 (ebook) | ISBN 9781499452709 library binding | ISBN 9781499452693 paperback | ISBN 9781499452716 ebook
Subjects: LCSH: Marine mammals–Juvenile literature | Fishes–Juvenile literature | Marine mammals–Anatomy–Juvenile literature | Fishes–Anatomy–Juvenile literature | Anatomy, Comparative–Juvenile literature
Classification: LCC QL713.2 .K64 2026 (print) | LCC QL713.2 (ebook) | DDC 597–dc23/eng/20250212
LC record available at https://lccn.loc.gov/2024061399
LC ebook record available at https://lccn.loc.gov/2024061400

Manufactured in the United States of America

Some of the images in this book illustrate individuals who are models. The depictions do not imply actual situations or events.

CPSIA Compliance Information: Batch #CSPK26. For Further Information contact Rosen Publishing at 1-800-237-9932.

Find us on

CONTENTS

OCEAN LIFE

Earth's oceans are filled with so many different fish. They come in many shapes and sizes, but they often have **similarities**. Many fish have fins. Many fish have scales too.

Did you know not all animals that look like fish are actually fish? Some are marine mammals! These animals are much like land mammals. For example, female marine mammals give birth to live young and feed them milk from their bodies. Let's use our skills of observation and **investigation** to learn more about fish and marine mammals.

DO YOU KNOW WHICH OF THESE ANIMALS IS A FISH AND WHICH IS A MARINE MAMMAL?

DEEP BREATHING

All animals need to breathe air. People have **organs** called lungs that take in a gas called oxygen from the air when we breathe in. Our bodies need oxygen to **maintain** life. Do fish or marine mammals have lungs?

All mammals, from people to dolphins and whales, have lungs! However, dolphins and whales don't breathe through their mouth like we do. They breathe through a hole on the top of their head. This is called a blowhole.

So, if fish don't have lungs, how do they breathe? Fish are animals, and all animals need oxygen. How do you think fish get oxygen from the water they live in?

SOME FISH HAVE GILLS BUT CAN ALSO BREATHE AIR LIKE PEOPLE! THEY HAVE A LUNG-LIKE ORGAN CALLED A **LABYRINTH**. THE POPULAR PET FISH CALLED THE SIAMESE FIGHTING FISH—OR BETTA FISH—IS A LABYRINTH FISH.

Many fish breathe through organs called gills. Gills usually look like **slits** in the side of their head. Water passes through the gills when fish swim. Gills take in oxygen that is **dissolved** in the water. Some fish can also take in oxygen through their skin.

FINS OR FLIPPERS?

Many fish and marine mammals use fins to move through the water. This is one reason they can look so similar. However, some marine mammals—such as seals, sea lions, and walruses—have flippers instead of fins. What's the difference?

Both fins and flippers are used to get around. However, a seal's flippers are more like legs and feet than they are like fins! In the water, seals use their flippers to move quickly and turn sharply. On land, seals use their flippers to hop along.

WARM OR COLD?

Warm-blooded animals have bodies that keep the same body heat no matter what the temperature is around them. Cold-blooded animals can't do that. When it comes to fish and marine mammals, are they are warm or cold blooded?

WALRUSES IN ICY WATER

All mammals are warm blooded. All fish are cold blooded. Fish use **adaptations** to stay warm. Many migrate, or move, to warmer water. Some are simply built for life in icy waters. Others bury themselves in mud to stay warm.

BABIES!

Like all mammals, baby marine animals grow inside their mother until they're ready to be born. Mother seals, dolphins, whales, and more all take care of their babies as they grow into adults. Do you think fish have live births too?

ORCA WITH BABY

Very few fish have live births. Many sharks give birth to fully formed baby sharks. However, sharks aren't known to care for their young. Most fish **reproduce** by laying thousands of eggs. Some fish will keep the eggs safe, while others leave them.

SEA SMARTS

Salmon are fish that live in the ocean. They swim up rivers to lay eggs in the same place where they were born. Blue whales use sounds to **communicate** over thousands of miles! Which do you think is smarter: fish or marine mammals?

SALMON

While some fish display **complex** actions, mammals are far smarter! Their brains are larger and more complex. They have many ways to communicate with one another. They also are known to form close bonds with family members.

SWIM MEET

We've learned about many of the **traits** that make marine mammals and fish different. But there's one trait that both of them share—most of them are excellent swimmers!

THE SLEEK SAILFISH CAN REACH A SPEED OF 68 MILES (109 KM) PER HOUR! HOWEVER, NOT ALL FISH ARE SLEEK SWIMMERS. PUFFERFISH ARE SLOW SWIMMERS.

Most fish have flat scales and slim bodies. is allows them to slide through the water. ost marine mammals have smooth skin and dy shapes. Fins and flippers are both great moving gracefully through water. Even polar ars, which spend half of their lives on land or , are excellent swimmers.

NOW YOU KNOW!

Fish and marine mammals have many key traits in common. So much so that it can be hard to tell some of them apart! However, we've learned about several adaptations and traits that help us to know the difference between the two.

There's so much more to learn about marine mammals and fish. And that's exciting! Just because this book is reaching an end doesn't mean that's all there is to say about these cool creatures. Keep on observing and investigating!

KNOWING WHAT YOU NOW KNOW, WHICH OF THESE ANIMALS IS A FISH, AND WHICH IS A MARINE MAMMAL?

GLOSSARY

adaptation: A change in a living thing that helps it live better in its habitat.

communicate: To share ideas and feelings through sounds and motions.

complex: Not easy to understand or explain; having many parts.

dissolve: To mix or cause to mix with a liquid so that the result is a liquid that is the same throughout.

investigation: To study by closely observing and asking questions.

labyrinth: A maze, or something shaped like a maze.

maintain: To keep something in good shape, or to carry on in the same way.

organ: A body part that does a certain task.

reproduce: To produce new individuals of the same kind.

slit: A narrow cut or opening.

similarity: A quality that two or more things have in common.

trait: A quality that makes one person or thing different from another.

FOR MORE INFORMATION

BOOKS

Salvatore, Kaitlyn. *Sea Mammals*. Buffalo, NY: Britannica Educational Publishing, 2025.

Scheffer, Janie. *Great White Sharks.* Minneapolis, MN: Bellwether Media, Inc., 2025.

WEBSITES

Learn About Marine Mammals
www.marinemammalcenter.org/animal-care/learn-about-marine-mammals
This fact-filled website has plenty of information about the different kinds of marine mammals.

National Geographic: Clown Anemonefish
kids.nationalgeographic.com/animals/fish/facts/clown-anemonefish
Read more about the clownfish and see up-close photographs of them.

Publisher's note to educators and parents: Our editors have carefully reviewed these websites to ensure that they are suitable for students. Many websites change frequently, however, and we cannot guarantee that a site's future contents will continue to meet our high standards of quality and educational value. Be advised that students should be closely supervised whenever they access the internet.

INDEX